W9-CZI-948

Tarantulas

Patrick Hely

PowerKiDS press™

New York

Published in 2018 by The Rosen Publishing Group, Inc.
29 East 21st Street, New York, NY 10010

First Edition

Editor: Melissa Raé Shofner
Book Design: Rachel Rising

Photo Credits: Cover, Joel Sartore/Getty Images; Cover, davemhuntphotography/Shutterstock.com; p. 4 (scorpion) FMPortella/Shutterstock.com; p. 4 (tick) Nataliia K/Shutterstock.com; p. 5 Photofusion/Universal Images Group/Getty Images; p. 7 Sebastian Janicki/Shutterstock.com; p. 8 pets in frames/Shutterstock.com; p. 9 urbazon/Shutterstock.com; p. 11 © iStockphotos.com/kimberrywood; p. 12 Stefan Zaklin/Getty Images News/Getty Images; pp. 13, 17 INTI OCON/AFP/Getty Images; p. 15 © iStockphotos.com/CathyKeifer; p.16 Coprid/Shutterstock.com; p. 18 © iStockphotos.com/Ableimages; p. 19 © iStockphotos.com/tiburonstudios; p. 20 somsak mungmee/Shutterstock.com; p. 21 © iStockphotos.com/the4js; p. 22 Dejan Stanisavljevic/Shutterstock.com.

Cataloging-in-Publication Data
Names: Hely, Patrick.
Title: Tarantulas / Patrick Hely.
Description: New York : PowerKids Press, 2018. | Series: Our weird pets | Includes index.
Identifiers: ISBN 9781508154198 (pbk.) | ISBN 9781508154136 (library bound) | ISBN 9781508154013 (6 pack)
Subjects: LCSH: Tarantulas as pets–Juvenile literature.
Classification: LCC SF459.T37 H45 2018 | DDC 639'.7–dc23

Manufactured in the United States of America

CPSIA Compliance Information: Batch #BS17PK: For Further Information contact Rosen Publishing, New York, New York at 1-800-237-9932

Contents

Meet the Tarantula

When many people see spiders, even the tiniest ones, they want to scream and run away! However, spiders can make good pets. Most pet spiders are safe and very simple to take care of.

Can you imagine having a pet tarantula? These giant, hairy spiders look scary, and they do have **venom**! In spite of their scary appearance and venomous bite, tarantulas are becoming more and more popular as pets. Is a tarantula the right pet for you? Let's find out.

scorpion

tick

PET FOOD FOR THOUGHT

Spiders are arachnids. Arachnids have two body sections and eight legs. Ticks and scorpions are arachnids, too.

Although tarantulas have venom, it's not usually strong enough to hurt a person. Tarantulas almost never bite owners who treat them well.

Creepy Crawlers

Many tarantulas are as big as your hand, and some are bigger. Their eight legs are long and thin, and they have tiny claws and hairs that help them climb. Tarantulas are covered with hairs that help them sense the world around them. This is important, because although tarantulas have eight eyes, they can't see very well.

Tarantulas have strong **jaws** and sharp **fangs** for catching the bugs they eat. These creatures sure do look scary!

PET FOOD FOR THOUGHT

The Goliath birdeater is the largest tarantula—and spider—in the world. It can grow to about 11 inches (27.9 cm) long and weigh up to 6 ounces (170.1 g).

There's a reason tarantulas appear in scary movies. They're creepy looking!

Tarantulas in the Wild

Tarantulas live in warm areas all over the world. Most live in South America, and none live in Antarctica. Like most spiders, tarantulas make silk in their body. However, tarantulas don't spin webs to catch **prey**. Some create silk mats near the ground. Others line their **burrow** with silk to make it easy to enter and exit.

Tarantulas are ambush hunters. That means they hide and surprise prey that wanders by. Bugs are a tarantula's favorite meal. Larger tarantulas may also eat frogs, toads, birds, and small **rodents**.

Female tarantulas can live for up to 30 years in the wild. This is something to keep in mind if you're thinking about getting a pet tarantula.

Choosing a Tarantula

There are more than 800 tarantula **species** in the world. In 2016, scientists named 14 new species in the United States alone. Do all tarantulas make good pets?

Just like different types of dogs, different tarantula species act differently. Some are more active. Some barely move at all. Young tarantulas are usually less expensive than adults, but they're harder to care for. Females live longer than males. It can take some time to choose the best type of tarantula for you and your home.

PET FOOD FOR THOUGHT

You may find tarantulas for sale in a local pet shop. Online pet stores ship tarantulas right to your door!

Tarantula owners know that these big, hairy spiders aren't scary or mean at all.

11

Tarantula Tanks

Tarantula tanks can be expensive, but they're necessary. Some tarantulas live in trees. They like a tall tank with branches to climb. Others live close to the ground. All tanks need a few inches of bedding for your spider to dig in. Always keep a lid on the tank so your tarantula won't escape.

The tank will need a heating pad for the spider to warm up on. Tarantulas aren't messy pets. You'll only need to clean the tank a few times a year.

Never keep two tarantulas in the same tank. They'll fight to the death!

13

Mmm, Crickets!

Keeping your tarantula well fed is easy—as long as you don't mind feeding it crickets! Tarantulas mostly eat small bugs, and crickets are their favorite. You can buy feeder crickets at a pet store. Most tarantulas eat only about two crickets a week. Larger tarantulas may eat up to six crickets a week.

It's important to "gut load" crickets before they're fed to a tarantula. This means the crickets have been well fed recently, too. Pet stores that sell crickets also sell cricket food.

PET FOOD FOR THOUGHT

Tarantulas can't eat solid foods. They pump a special **enzyme** made in their body into their prey. This enzyme turns the prey to liquid, and the tarantula sucks up its meal!

Your pet tarantula will hunt down the crickets you feed it. You should also provide a shallow bowl of water so your tarantula can drink.

15

Illnesses and Injuries

Dehydration is a common problem for tarantulas. This means they aren't getting enough water. A wrinkled, sick-looking body is a sure sign of dehydration.

Like many spiders, tarantulas **molt** as they grow. They need to lie on their back and may look dead. After molting, the spider's body is soft and tender. Never bother a tarantula when it's molting. Tarantulas can experience a "bad molt." This can result in cuts and lost legs. However, tarantulas can regrow lost legs!

PET FOOD FOR THOUGHT

Many owners use superglue to help heal cuts on tarantulas.

Tarantulas that are well cared for don't often get injured, or hurt, but it can happen. Tarantulas can be injured when they fall from a height or if you don't handle them carefully.

Be Careful

Pet tarantulas rarely bite their owners, especially if they're treated properly. When tarantulas do bite, it feels like a bee sting. Some people might be **allergic** to tarantula venom. Owning a tarantula can be dangerous, or unsafe, for these people.

Some kinds of tarantulas can "throw" body hairs when they sense trouble. These hairs may stick in the eyes and nose of predators. They may cause **itching** and pain. It's a good idea to wear safety glasses when handling a pet tarantula.

People who are allergic to tarantula venom may experience itching, hives, or shortness of breath if they're bitten. They may also feel **nauseous**. Doctors have drugs to treat tarantula bites.

19

Pets Aren't Wild

Some people release tarantulas into the wild if they feel they can't care for them anymore. This is dangerous for the tarantula as well as for other animals in the area. A tarantula may not be prepared to survive where you live. It might be too dry or too cold. They may eat too many native creatures and harm their populations, which in turn can harm other plants and animals.

Tarantulas might seem like a weird pet, but they can also be fun!

Are you brave enough to keep a big spider as a pet?

Tarantula Care Fact Sheet

- Tarantulas need a special tank to live in.

- Give your tarantula bedding to dig in.

- Tarantulas love to eat crickets.

- Moths, cockroaches, and other small bugs also make tasty snacks.

- Make sure to place a shallow bowl of water in the tank.

- Always keep the lid on your tarantula's tank.

- A heating pad will help keep your tarantula warm.

- Keep the tank away from bright lights. Tarantulas like the dark.

Glossary

allergic: Having a bad bodily reaction to certain foods, animals, or surroundings.

burrow: A hole an animal digs in the ground for shelter.

enzyme: Matter made by cells that causes changes to other matter.

fangs: Sharp, hollow, or grooved teeth that inject venom.

itch: An uneasy feeling on the surface of the skin.

jaw: Either of the two bony parts of the face where teeth grow.

molt: To shed hair, feathers, shell, horns, or skin.

nauseous: Feeling as if you are going to throw up.

prey: An animal hunted by other animals for food.

rodent: A small, furry animal with large front teeth, such as a mouse or rat.

species: A group of plants or animals that are all the same kind.

venom: A type of poison made by an animal and passed to another animal by a sting or bite.

Index

Websites

Due to the changing nature of Internet links, PowerKids Press has developed an online list of websites related to the subject of this book. This site is updated regularly. Please use this link to access the list: www.powerkidslinks.com/owp/taran